A Chipmunk on my Shoulder

A Chipmunk on my Shoulder

G J Helbemae

ASHFORD
Southampton

First published in 1969 by Angus & Robertson Ltd

This edition published in 1989 by Ashford, 1 Church Road,
Shedfield, Hampshire SO3 2HW

British Library Cataloguing in Publication Data

Helbemae, G. J.
  A chipmunk on my shoulder. – 2nd ed.
  1. Pets. Chipmunks. Biographies
  I. Title
  636'.93232

  ISBN 1–85253–200–9

Phototypeset by Input Typesetting Ltd, London
Printed by Hartnolls Limited, Bodmin, Cornwall, England

Dedication

When as a grown man I was at last able to fulfil my boyhood dream of having squirrels as pets, I had no thought of using them as subjects of my next book. While observing these "sparks of life" in my flat, however, I realized that the book was growing of itself like the sunflower seed which my chipmunks love to bury in our plant pots.

Therefore this book is dedicated to my little friends — the chipmunks — not only for having encouraged me to write about themselves but for remaining delightful companions throughout our long association together.

Contents

Foreword

In his native Estonia G J Helbemae kept a
bewildering variety of pets, from white mice to
leeches, but the occupation of his country by
the Soviet Red Army in 1944, and his sub-
sequent exile in England, meant that it was
many years before his thoughts again turned
to the animal he admired most — the squirrel.
Having established himself as a respected
writer in London, producing many radio and
children's plays and a number of well-received
novels in Estonian, it was not until a summer
holiday in the Black Forest, surrounded by
squirrels, that his childhood ambition to keep
one was reawakened. Even then it was not the
native red or the immigrant grey squirrel which
he finally tracked down in a seaside pet shop,
but the multi-striped Asiatic Tree Squirrel —
more commonly known as the chipmunk.

Foreword

The first successful arrival, the even-tempered Miku, had not long settled into the sofa of the author's London flat before his territory was rudely invaded by the mercurial Minni, whose disruptive presence soon necessitated a complete reorganisation of the author's domestic life. Clothes and tissues had to be sacrificed for nest-building, cupboards and windows kept shut for fear of petty larceny and unwanted escape, and valuable working hours given up for feeding and observation. But the effort was richly rewarded by the friendship and trust of these previously unfathomable creatures, these "sparks of life" as the author referred to them. He learned to interpret their behaviour quite accurately; to know when Minni was terrorising Miku or merely teasing, when she was feigning hurt or genuinely upset.

Minni and her less adventurous companion Miku brought, and will continue to bring, joy and illumination to thousands via the author's sensitive and appealing account of their frolics and foibles. We feel a real sadness and desperation, mirroring the author's own, when he is at last driven to comb the parks of London for his beloved and irreplaceable companion.

CHAPTER ONE

Living with chipmunks

It is after nine o'clock in the morning.

I am sitting at my desk opening the mail, but now and then my eyes stray towards a large wardrobe and to the cupboard which stands on top of it, reaching all the way up to the ceiling. One of the doors is slightly ajar and it is to this opening that my glance keeps straying: to see whether she has woken up yet. "She" being my friend Minni, the chipmunk.

Minni's first choice of nest was in the chest of drawers in the hall, in which she had discovered a back passage. In one of the drawers she found a piece of silk material, gnawed a few air holes into it and slept like a princess — wrapped entirely in silk.

Soon after, however, she changed her mind and moved into my study, into the top cupboard where we keep suitcases and boxes of

1

clothing. One of the boxes contained my swimming trunks and my wife's swimsuit. Minni laid claim to this box, chewed a large hole into the lid and made her nest inside my trunks. She had of course arranged the rest of the "furniture" in the box to her liking, making herself an air vent and in the autumn lining the nest with tissue handkerchiefs filched from my wife's dressing table. These were chewed into tiny pieces.

One of the chief reasons for this removal was that we had kept opening the drawers in the hall chest to look for things, thus trespassing on her property.

Minni is still asleep. O.K. Time to get down to my work, which is concerned with an Estonian-language newspaper published in London. I put a clean sheet of paper in the typewriter and begin. I type two letters when something jumps out of the typewriter and into my face. A nut! Minni must have hidden it there last night. She is always up to such tricks when I am not in here. Probably trying to tease me?

It has happened that I have had to leave the room for a moment and on my return when sitting down again, I have had to rise with lightning speed. Minni during my absence has put a hard sharp nut on the chair. No harm when it is only a nut, but there was a time when it was a raspberry and my trousers are still stained to this day. I now inspect my chair before sitting down, especially if fresh fruit has been a part of her menu.

Eventually I manage to type a page of foolscap and as Minni has still shown no sign of life, I decide to call her:

"Minni, Minni . . ."

A fierce rustling starts in the top cupboard and Minni's sleepy face appears at the door. Then she crawls to the edge of the cupboard and looks over the top. It is the same every morning. After a full minute of "pulling herself together" she jumps down to the topmost shelf of the bookcase — which extends the whole length of one wall — so that she can run along the books straight to my desk, remaining just above it. There she will stop and look at me as if to say:

"Good morning! Here I am."

"You are very lazy this morning, Minni," I tell her.

"Tsk, tsk," she replies noncommittally and starts on her morning toilet, first wetting her front paws with her tongue and then scrubbing her cheeks, nose, eyelids, ears and the back of the ears as far as her paws will reach. Next the

3

tail. After all, the tail is a squirrel's pride! The tail is pulled forward between her legs and is literally combed through while the front paws guide it past her clipping teeth.

But the teeth themselves haven't been cleaned yet. For this Minni uses a peach or plum stone as a toothbrush, lifting it between her front paws and chewing it. This irregular type of stone-chewing is important to a squirrel's teeth; it keeps them in good condition and stops them growing too long. Sometimes it may happen that one of Minni's "toothbrushes" falls from her mouth and on to my desk. She watches with interest as I pick up the stone and hold it in my fingers. However, she shows me that she has plenty of "toothbrushes" to spare and simply chooses another to finish the job.

After the toilet is completed, Minni jumps from the bookshelf on to the curtain and begins to make her descent. This is the first part of her morning exercises, which she does by hanging on to the curtain with her back legs and stretching herself as far as she can with her front paws hanging downwards: swinging there like a bat. This is repeated many times, sometimes accompanied by a wide yawn. As Minni always does her exercises in the same

4

spot every morning, the place in the curtain where she hangs has been worn pretty thin.

Now the final traces of sleep have been swept away and Minni leaves the curtain to perch on top of the radio, which is to my left and within reach.

From here we observe each other. Minni's sharp black eyes follow my every move; sometimes when I blink, she blinks too.

"How about some nuts, giant?" she seems to ask me.

I think Minni regards me as a large beast. When she was still three months old and had just left a nest of five, and prior to our friendship, she stopped in front of me one day and began to study me slowly from the soles of my feet to the top of my head, and as her eyes were forced to travel higher and higher so her body stretched too until she stood before me like a ballerina on the very tips of her toes.

"My, what a large beast you are — but there's no need for fear," seemed to be the conclusion she reached, for she soon started to come to me freely, eating from my hand and jumping on to my chest from quite a distance, or simply climbing up my body by the way of my trouser legs.

So now, in answer to Minni's questioning

look, all I have to do is to reach for my pocket and she is instantly on my shoulder . . . and the next moment on my hand. I hold the cobnut lightly in my half-closed palm. Minni pushes her nose down to take the nut, and will then remain on my hand to crack it or jump to the edge of my desk. Cracking the nut only takes half a second. If she intends to eat the kernel right away, she retains one half of the nut shell, holding it between her paws like a plate, balancing the nut on it. Having eaten half, she pushes the rest into her cheeks and asks for more. Minni uses her cheeks as "shopping baskets" in which she collects her food; each cheek is big enough for three or four average-sized nuts, or one spoonful of corn or sunflower seed. As she is asking for more for her "baskets" I give her two more nuts, one for each cheek. In the morning this is usually enough, and it sends her jumping joyfully, her cheeks "mumpsy", into the hallway.

The chipmunks' dining table is in the hall of the flat, but this isn't where she is heading. She is on her way to the kitchen — a much more interesting place, where the unexpected may be found. Soon after I hear a loud clattering of pots and pans. It is all right, she is only inspecting the breakfast dishes, stacked next to the sink; putting her head inside the coffee cups to lap at the hardened sugar. The cup or a spoon falls. A guest once commented on the noise Minni makes in the kitchen, saying a huge cat is so much quieter. And yet for all the noise, she has only broken one cup — so far!

One day I heard a definite shatter follow a crash. I ran to look. There was Minni sitting on the floor, in the middle of the broken pieces, wagging her tail like a dog. Yes, she even wags her tail, though a little slower and usually out of excitement, anger or fear — not joy or well-being. What had happened was that we had left the door of the china cupboard open and Minni had immediately gone to investigate. It looked as though she must have jumped straight into a cup and fallen down with it. She was still sitting in the broken porcelain base, wagging her tail and looking at me perplexed with a "And how do you think this happened?" expression. I was worried in case she had hurt herself, but as soon as I reached for the bag of nuts (from a place which was out of her reach) she revived immediately, jumped out of the debris and on to my hand, and was soon eating as though nothing had happened.

Most of the troubles that my wife and I experience arise from our own forgetfulness —

that there are chipmunks in the house. For instance, I may leave a cake uncovered on a table, or a plate of biscuits. Minni nibbles them all through without really wanting them — but as they are there, might as well! She will start then to drag them from the table, up the curtain to the top of the kitchen cupboard. This cupboard too constitutes a part of her chain of storerooms. Some biscuits are so big they are half her size, like millstones, but she manages them all in time, if she isn't interrupted. It is the same with the fruit dish; if there happens to be a half apple on it, she soon cleans it of all its pips. Nothing tastier. And the grapes there are definitely sweeter than the grapes on her own table . . .

As soon as the noises in the kitchen die down, I know that Minni has finished her inspection and gone into the living room.

Meanwhile I have managed to get on with my work and finished several columns for the paper, the telephone has rung, I have spoken to a number of people, and I am about to put another sheet of paper into the typewriter when suddenly — a charge of cavalry comes galloping into the room, the curtains shake as if in a storm, and there is one flash of striped lighting followed by another. For an instant my desk is a battlefield. Papers fly into the air. The galloping pursuit continues out into the corridor, grows fainter and then — silence.

What is going on? The charge of the Light Brigade?

No. It is my other chipmunk, Miku, who resides in the living room in a space under the cushions of the settee, and who must have been in the middle of his exercises when Minni disturbed him. Within a week of her arrival, Minni made it quite clear to the older but softer-hearted Miku that the curtained part of each room belonged to her, and her alone. And that only she, Minni, had the right to do her callisthenics hanging on to those curtains, climbing up and down them and looking out of the window. Miku was assigned the other half of the room as his territory.

I am afraid that Minni, true to her character, never consulted Miku about these arrangements and Miku just couldn't understand why the newcomer should claim the best part of each room for herself. So he resisted, completely ignoring Minni's rules.

Minni, however, is very egotistical and selfish. She will have her own way and if Miku won't co-operate, she'll soon show him! So it happens every day that whenever Miku goes

near the curtains in any of the rooms, Minni
rushes straight into battle and with such deter-
mination that Miku flees in fear. At first she
would leave Miku alone if he had already fled
from the curtains to the other side of the room.
But as he persisted in trespassing at every
available opportunity, Minni decided to step
up the campaign and to chase him from room
to room until he was forced to come to me for
refuge. Sometimes I have been able to put my
hand between the two of them. Minni has
jumped into my palm in the heat of the battle,
but suspecting a trap has jumped out again to
continue the chase. Because of their speed it is
not easy for me to get my hand between them
to intervene. The chase usually ends with Miku
retreating to his nest inside the settee — where
not even Minni will follow. Even an uppity
chipmunk maiden like Minni does have some
sense of propriety — not to follow a gentleman
into his bedroom!

Today Minni hasn't the energy to chase Miku
to his nest. She remains at the door, listening
for sounds from the living room, and being
satisfied, comes back to my desk. She sits on
the back of my chair, and throws me a look of
genuine complicity as if to say, "Well, we got
rid of him, didn't we!"

"Minni, that wasn't a nice thing to do," I try to tell her.

"Hah! He's just a dumb bell, that's all," is her answer to that

I have finished typing a news item at last and as I remove the sheet of paper from the typewriter, I see from the corner of my eye that Minni is still squatting on the back rest of my chair, apparently deep in thought but actually waiting. I know what she is waiting for. All I have to do is to reach for a red-topped box, which rattles invitingly full of sunflower seeds and grains of corn. Like lightning she appears on my hand, and even before I can open the lid, she impatiently assists me with her nose.

But other times she will veer away and retreat to the desk, looking at me with reproach and disgust.

"Take the smelly old twig from your mouth!"

Of course. I have forgotten to remove my pipe. As soon as I put it away she returns to my hand and peers inside the box. If she doesn't find a peeled nut among the contents, she shows her disappointment by going back to the desk or chair, to sit there with her tail towards me, sulking. It is her way of saying:

"I want more nuts. This here doesn't interest me."

11

However, in the event of the sulks not working, she returns, sometimes via my pockets, so that only her tail stays visible, and rattles about among the loose change hoping to find a forgotten nut. When the search through the pockets proves fruitless, she returns once more to my hand, which I have now filled with some of the contents of the box. Resigned, she sets about the sunflower seeds. She holds the seed between her paws, and I hear a quick crunch and half the shell flies out from one side of her mouth. The other half is still held tightly. She removes the soft seed with her teeth and then discards the rest of the shell. All this lasts a few seconds. So now Minni picks and chooses her seeds and only a few are eaten; the rest travel to her cheeks, always one to one side, the next to the other side. She doesn't stop until both shopping baskets are full to the brim. Then she carries them away quickly, along the same route by which she came, across the bookshelves, to her nest. Or sometimes just as far as the space behind the nearest book.

Having discarded her "mumps", she returns for more. If I am busy, however, and do not have time, I place the seed box next to my typewriter, and she then has to overcome the noise of the typewriter, by flattening her ears close to her head. A pity we can't do the same in order to escape the increasing noise in our cities!

Last of all, Minni gets to the corn. Cleaning the grains she eats and stacks them into her "baskets". She always takes precautions against the sharp ends of corn grains; she simply bites them off, so they won't harm the inside of her mouth.

When the box is empty and Minni is still on her "shopping spree", she just will not believe that the shop is closed. So she again returns to rout out my pockets, this time even taking her tail with her; but the picture is as grim as before and a disappointed little face soon appears from the slit. If our eyes meet, she scrambles out to my hand once more, sniffing and inspecting it, but if I am hard at work and haven't the time to meet her questioning glances, she climbs to the back of my hand and gives me a gentle reminder by biting at a finger or at the hand itself.

"Take notice of me, you big beast — I'm still here, you know."

In all honesty, I then show her both empty hands. There really isn't any more.

Minni thereupon resorts to some crafty ruses of her own. She runs to the door, stops there on her back legs and looks at me invitingly. I know from experience that should I answer her call to follow, she will run ahead into the kitchen and will stop on the chair which is directly in front of the cupboard door where we keep her foodstuffs.

"There's plenty there. All you have to do is to get it out," she seems to tell me. She knows because she has seen me taking nuts and corn from there before. One afternoon she even managed to put one of my momentary distractions of attention to good use. She jumped into the cupboard and got shut in for a whole night.

The next day, when I had already begun to worry, wondering where she had got to, I heard a great scraping and crying coming from the direction of the food cupboard.

"Seems that food isn't all in life and one's hunger soon gets satisfied," was surely Minni's experience during that long night.

I have to admit that sometimes I do get a bit soft-hearted and reach for another handful of sunflower seeds. Seeing this, Minni gallops back to my desk and jumps on the back of my chair — and as I approach she launches herself at my chest from a distance of three or four paces. Sometimes she jumps from as far as five or six paces, and breaks her fall by clutching

14

at my trouser leg. If she falls to the ground, she climbs up the leg, back to my hand.

But I am very busy today. I have to harden my heart. Minni is still sniffing round the empty box. She gives it a final disdainful push with her nose so that it falls against the typewriter, and disappears . . .

The printers ring me up. They give me an hour before they want the articles I have just typed. I decide to have a quick lunch and go to the kitchen to warm up some fresh vegetable soup containing peas from the pod. I hardly have time to drink the first spoonful when the phone rings. I answer it in the living room, on the extension. Out of the corner of my eye I see Minni sitting on the windowsill, looking out of the window with rapt concentration. She comes to me presently, while I am still in conversation on the phone, to see if by chance I have put a few nuts into my pockets or into the palm of my hand. Alas no! So as I am making notes, she bites the pencil and then goes back to the windowsill. The call over, the notes taken, I return to my now cold soup. The first thing I see is a guilty-faced Miku making a quick retreat from my plate, his long tail erect as he jumps from the table to the floor and flees . . .

15

I look at my plate; it is rimmed with empty outer pea skins. It seems as if the usually shy and retiring Miku has eaten up all the peas from my soup by peeling away the tender outer skins and sucking the contents. Needless to say the chipmunks' own table was already richly heaped with fresh peas from the pod. But no doubt they taste much better when stolen out of my soup!

I have to dash to the printers.

Minni and Miku will stay at home alone. Perhaps for a number of hours. I know there will be some surprises awaiting me when I get back. But after all, it was of our own free will that we chose to live with chipmunks.

CHAPTER TWO

Miku and Minni

I have often been asked:

"How come you chose chipmunks as pets?"

I think I can answer this quite shortly. As a child in Estonia I had all kinds of unusual pets, among them lizards, leeches and white mice. Most of all I craved a squirrel, but never did get one.

However, a few years ago, while spending a summer holiday in the Black Forest, I saw lots of squirrels; one even came and ate out of my hand, and I became completely enraptured. My childhood desire was reawakened. "I want a squirrel!" I said to myself. On returning to my home in London, I went into one of the largest and best-known pet shops in a department store, and there I saw a number of striped Japanese squirrels. They were, however, terribly expensive, and I was not too keen on their

17

tails, which reminded me of rats' tails; but later I met a friend who told me that he had seen some lovely little striped squirrels in a small English seaside resort.

I travelled there and returned with a chipmunk. My childhood dream had been fulfilled — forty years later.

This was Miku the First, and unfortunately it transpired that I had been misled, having been told at the pet shop that I had a young chipmunk, who later turned out to be old and sick, and who had been looked after poorly. All the same, we became fairly good friends, especially when I myself was taken sick and Miku, despite it being winter and really time for his hibernation, decided to stay by my bedside and keep me company.

As it happened, Miku the First was still alive when I purchased my second chipmunk, aged six months, in London. This was supposed to be a young female, but, she turned out to be a young male. The shop was willing to exchange him but he had already been with us a few days and I had become very attached to him. He was extremely kind towards his sick companion and stopped to lick him from top to toe frequently during the day. He also held his head on one side for me to tickle him under

18

his chin, and when I spoke to him he had a habit of moving his lips, as though in answer.

Miku the First died and Miku the Second was left alone. I decided to get him another companion, but this time very definitely a young lady. I went back to the London shop where the mistake had occurred, and bought a female. On putting "her" into a carrier, I expressed some doubt, and when we made another inspection we ascertained that they had confused the sex yet once more.

As it happened, they only had one other chipmunk in the shop, which according to them was female. We never managed to check her credentials, however, for she made a quick getaway through a hole in the floor to a store-room downstairs. For a whole week they chased her, with butterfly nets, but they could not catch her — or him, or it.

Unfortunately there were no more chip-munks to be had in London, so I phoned the seaside pet shop (the one which had cheated me over Miku the First), and they informed me that they had a whole family of five chip-munks — only three months old. And yes, there were females among the litter. They would send me one to London. Which they did.

19

I met the train and claimed a wired cage full of straw. In the straw somewhere, presumably, lay my female chipmunk. At last!

Arriving home, I carefully emptied the whole contents of the wired box into a bigger cage that I had. I heard a soft thud among the straw and laid eyes upon my new boarder for the first time. The sight of her was quite a shock. Something was not right with her appearance. Compared to Miku, she was a mass of stripes. I began to count them — there were five in all. I counted again. Still five. Miku only had four stripes, and so for the first time in my life I saw a chipmunk with five stripes.

I had read the Red Indian fairytale of how the chipmunks got their stripes. In the beginning chipmunks were reputed to have been of a plain colour, a brownish grey; but during a meeting of all the animals of the forest it happened that a bear and a chipmunk got into an argument. The bear became so angry that he lashed out with his paw. The chipmunk was too quick for him and jumped out of the way, with the result that the bear's claws instead of killing him outright, left furrows on his back — which are now his stripes.

Well, I thought, if it is five, it is five. But what about Miku? How will he react to a five-striped chipmunk? He was from the West Coast of America, while the five-striped chipmunk which we had now acquired was from the East. I had read that there are more than sixty varieties of chipmunks in the squirrel family to which they belong, and which itself has well over three hundred species.

Miku's reaction was one of considerable curiosity. Being free himself, he was able to watch the newcomer from any vantage point he chose, which he did with ardent nosiness, while washing his face in preparation for the inevitable meeting.

I opened the cage and extended an exploratory hand to my new friend. She was, however, frightened and set up such a high-pitched squeaking that I pulled my hand back quickly. "Minni, Minni," I said soothingly, whispering her new name. (Which had been Miku's for a while, until his true sex was discovered!)

Minni eventually, while washing her face too, stared back at me with bright eyes and flirtatious looks. Seeing this, I was convinced that my new tenant was indeed a full-blooded female, even if she did have five stripes!

I left the door of the cage open and went to sit elsewhere to await events. Miku was circling the cage and no doubt encouraging Minni, who

21

soon came out. They stopped and confronted each other and, rising on to their hind legs, rubbed noses. Perhaps a proper formal greeting, but then, perhaps not — for they immediately jumped apart. Minni returned to her cage, sat there for a moment to recover, and then came out again. The two of them then met in the middle of the room for a game, much like ring-a-ring-a-roses.

Soon Minni's interest in her surroundings grew as her confidence increased. She discovered the joys of climbing up the curtains and sitting on the pelmet — a safe place. But she did not remain there for long. There was so much else to be seen in the flat, and she dashed off like a flash.

The speed of chipmunks is indeed amazing. Once, when both Miku and Minni were in the same room with me, close together, one chewing sunflower seeds and the other a piece of apple, I happened to sneeze. It was a matter of a fraction of a second, but all I saw were two "streaks of lightning" as one disappeared behind the curtain and the other under the bookcase. I had never seen such speed in such small animals and began to wonder whether they hadn't broken the sound barrier and done themselves inestimable injury? Half a minute later, however, they emerged quite calmly as though nothing had happened.

I noticed right away how much bolder Minni was than Miku had ever been, and how much more alert and curious. In only a matter of hours, she allowed herself to be looked at from close range, summing me up before returning to the sanctuary of the pelmet. There was a definite, noticeable, difference between the two types of chipmunk. Minni's tail was one inch shorter than Miku's and her back legs were just a bit shorter too.

As to their eating habits . . . Miku drank milk silently, like a well-brought-up child. Minni, however, slurped and burped so that the whole flat echoed. Miku would then look at me with a shamefaced apologetic look, as if to say:

"Oh dear! What manners! And she's supposed to be a lady!"

Miku's manners are impeccable. He even wipes his mouth with a napkin after each meal. Actually it is the corner of the tablecloth on the chipmunks' eating table — but one can't be too critical! Minni somewhat redeems herself by washing her face after each course; she eats sweet corn — washes her face; eats sunflower

23

seeds — washes; eats fruit — then washes again.

Despite their differences, I hoped that Miku and Minni would soon get to know each other and form a friendship. Their initial meeting had been successful and all throughout the first week there were plenty of encouraging signs. For instance, the first night, Minni went to sleep in the same curtain fold as Miku. But a few days later she sought her own bed, choosing a drawer in the chest of drawers we had in the hall. Miku went to sleep in the settee, where Miku the First had slept.

There was no real enmity to be noticed — they even played together during the day — until Minni decided to divide the apartment into East and West. I have already explained this in the first chapter, the dilemma it caused Miku, and how hard it was for him to abide by the new rules.

In their chases, I noticed that Miku no longer took Minni's pursuit as play. He was obviously a little afraid that she really would sink her teeth into his back legs, if he allowed her to, and as soon as she got too close for comfort he would jump straight up into the air — letting an unprepared Minni rush on headlong beneath him, sometimes even a distance of

many yards before being able to come to a halt.
By this time, Miku would be well away.

That Minni would bite him in the hind legs
seems to have been an inherent fear with him
and common to squirrels in general, there
being certain species where hind leg biting is
a form of fighting, the goal of which is the
severing of the tendon — this making it
impossible for the enemy to jump. In America,
the small red-brown squirrels are known to
torment the large grey squirrels in this manner,
and the larger grey lives in fear of its smaller
brother. It seems that squirrels actually do
possess an Achilles heel!

When I realised that their chases were no
longer friendly romps but territorial per-
secutions, I too began to worry about Miku's
heels. Minni, despite her own shorter hind
legs, was a very fleet and cunning young lady.
By far the worse aggressor of the two as well
as the faster.

"What if Miku doesn't jump away in time,
what then?" I used to wonder.

And I was sorry for him too. It was no fun
for him, having his home taken over by a bully.
My fears, however, were slightly relaxed when
I noticed that Minni, having caught up with
Miku and gaining the opportunity to bite,

restrained herself enough to make her goal rather one of getting in front of him and cutting off his retreat to his nest. Though her games were somewhat cruel, I was relieved to see they *were* still *games* to some degree, whatever their motive.

However, in the case of one particular curtain, Minni made an exception in allowing Miku to sun himself on top of the chest of drawers in the hall, under the window. I decided to move the chipmunks' dining table to the same spot (it had been in the living room till now), marking the dividing line between East and West.

At first this arrangement suited both of them and didn't lead to any trouble, other than the usual one of who would get the larger nuts and who the small ones. The large Brazil nuts had always been Miku's favourites; he would carry them to the windowsill in the summer, and to the front of the fire in the winter, chewing at them in the way a dog gnaws a bone. He didn't often finish them at one go, but now as soon as a half-eaten nut was left unguarded, Minni took it away immediately in her "shopping baskets" to a place behind the books or wherever she thought best. If I put a new nut in its place, it was usually Minni again who arrived first and took it from under Miku's nose. And Minni didn't even like those nuts, but she took them away, because they were Miku's favourites and good enough for her lair. I thought up a plan — for Miku's sake — and fastened the large Brazil nuts to a chunk of wood on the dining table with a piece of wire. Minni tugged away in anger, her tail wagging indignantly, but to no avail.

There was however, more trouble over cherries. I would put four on their table, two for each of them, as I knew from experience that two was enough. If I gave them more, they would perform as chipmunks do when confronted with plenty in the wild by nibbling first at one and then at another, leaving an unsightly mess of mangled fruit.

Minni would arrive immediately and eat her two cherries, including the stones which she cracked and emptied.

The other two, however, would be giving her a headache. She didn't really want them, but suppose Miku was to come? It was not her habit to carry away fresh fruit, but suppose she hid it? She would pick up the cherries and carry them to the top of the sash-cord window frame, rolling them into a corner so that they could not be seen.

27

Miku would arrive at the table, eat some sweet corn and a few sunflower seeds, all the while sniffing and smelling at the cracked cherry stones. Then he would raise his nose into the air and follow the scent — right up to the window frame. Ah! the two hidden cherries. He would eat them up right away, but leaving the cherry stones. (He doesn't like cracking nuts.) Having eaten, he would return to the dining table for a drink.

Minni comes up quickly — her thoughts still on the cherries. I hold my breath. Minni looks round the table. Then climbs to the window frame, where she left the cherries, to be confronted by two bare pips. She examines them closely, to be sure there is no mistake — but there is none. It is all quite clear. She turns round, pulling her ears flat against her head — her body thus taking the shape of a rocket, but her target is not the moon outside the window, but the unsuspecting, still quietly dining Miku. She pauses for a moment, takes aim and shoots down. But does not land on Miku after all. It seems that his sixth sense has warned him, for although apparently at ease, he was quickly alerted and jumped clear. Minni is left on the edge of the table, irate, but does not bother to follow — what is gone, is gone! Instead, she returns to the window ledge, in fact to the cherry stones, which she starts cracking as loudly as possible. Then she goes back to the dining room table, gathers up as much food as she can and takes it to her nest, making return journeys until the table is empty. Empty, that is, except for a piece of half mouldy apple and the chained Brazil nut.

"That's all you'll get now," she seems to be saying. "Now come and get what remains."

CHAPTER THREE

Making friends

It was unfortunate, but due to Minni's strong personality, my own relationship with Miku began to suffer. Being shyer and less extrovert, he required more attention, which however was constantly being stolen by Minni.

So whenever Minni was not near, I called Miku to my hand, offering him nuts. It took a while. Miku was showing willing but his timidity was still such that Minni was always on the scene before him, jumping on to my knee and facing Miku with the usual "rocket" head that said, "Just you try to come nearer" — and of course Miku fled.

Sometimes, when I would be quietly reading, Miku would suddenly jump into my lap; but Minni noticed this immediately, even from the other side of the room, when she was not even looking our way, and would pounce on Miku.

Do not think that my friendship with Minni happened overnight. Far from it. Minni was suspicious of my hand at first, especially if I reached out for her. This was probably a reflex derived from her pet shop days, when she was taken out of her nest comprising of a family of five and put in a box, and then endured a harassing journey to London. Now, although she came to collect nuts from around my hand, she still kept a wary eye on the hand itself. The slightest finger movement would be enough to send her scurrying.

The next step, of course, was to get her to eat from my hand, which she first started doing by "pinching" the nuts from me. At this stage, the hand had to be absolutely still.

Some weeks after that, however, I was at liberty to move my fingers gently. This didn't actually alarm her but made her fill her cheeks quicker and disappear sooner.

I did not really count the "eating from my hand" stage as true friendship. After all, what kind of friends are we if one is still suspicious of the other? Many like to think that once an animal has eaten out of your hand, it is your friend. But a squirrel knows its own swiftness

31

and is aware of every movement, and it is more a daredevil game of chance than trust. Just try moving the other hand!

It was not until Minni stopped cramming her cheeks full of food and running away, but stayed on my hand to peel her sunflower seeds, that I began to feel a mutual trust. She even stopped eating and looked at me sometimes, and soon learnt that the other hand was not there to grab and imprison her, but to stroke and tickle her chin.

Soon she let me brush her, and twist her tail around my finger, while I held her with my other hand. She stayed content and showed no sign of fear. Sometimes, when I was holding her, she would sing to me. A kind of chirping, which had doubtless originated the name *chipmunk*.

It was only at this juncture that I could really say we were really friends. Our friendship, however, introduced new hazards. For instance, before, there was no need to worry about little chipmunks getting under your feet. They were too busy fleeing out of the way, instinctively keeping their distance. Now Miku still stayed out of range, but not Minni. Whenever I met her in the hall, she either jumped up to my chest, which was all right, or stopped still, waiting for me to pass her, or worse still, would make a dash in my direction, expecting me to sidestep. A collision was imminent, especially if I was in a hurry. And such a collision actually took place once.

It was a Saturday, and my wife was vacuuming the hall carpet (the noise sent Minni to the pelmets in fright at first, but she soon got used to that). I was coming from my study in a great hurry. Minni ran out from the front room, and jumped out of the way of the vacuum cleaner straight into my path. I did try to stop on half a step, but my heel was down and the toe of my shoe was raised. It was against the toe that Minni stormed.

Poor Minni! But she was very brave about the whole thing. She didn't make a sound. Later, when offering her some nuts, I noticed that part of the fur on her cheek appeared to have been shaved off. It was in fact a bald swollen patch. That must have been where she met the full impact of my shoe. The swelling soon subsided and a few weeks later the fur grew back again. The whole incident was forgotten.

Minni loves to explore my briefcase but jumps out as soon as I lift it up. In this way I haven't made the unfortunate mistake of carrying her with me to the printers, or into town. She also likes other people's briefcases and shopping baskets, but gets out as soon as the owner approaches.

Often, on reaching for my coat, I have met the bright beady eyes of Minni, staring over the edge of my pocket. She has always jumped out, however, though leaving me a few sunflower seeds — to chew as I go along, presumably.

Minni is very sensitive to my feelings and despite our friendship, has a way of reacting to my moods. Should I be in a hurry, already in hat and coat, but have to return to my desk for a moment to get something, Minni scuttles away from me in terror, uttering shrill cries, and takes refuge behind the bookshelf. A soothing "Minni, Minni" will bring her back, when she will look at me with wonder: "is that really you?" and then verifying "it is" before returning to normal.

It also worries her when I carry large parcels. As far as she is concerned, it could be anybody coming through that front door, so she retires to the pelmets to see if the person behind the parcels is friend or foe. During the first months, Minni even became alarmed when I changed my suit or put on another coloured cardigan.

One day, while I was away for a few days on business, my wife decided to put on one of my suits, the casual suit I usually wore around the house, to see whether Minni could be cajoled into the same kind of friendship I had with her. Minni gave her one sharp look, snatched a nut from her hand and fled.

But Minni likes my wife's hair. When she is lying down quietly, Minni will creep up and start to play with it, picking up the strands with her paws as though trying to make plaits. If my wife remains motionless, the game continues, but should there be a sudden painful tug to which my wife cannot help but react, Minni runs away immediately.

Minni's acceptance of strangers depends on her sixth sense and instinct. She might even accept a nut from one while not even coming near another. Just one look-over usually sums it up — either the stranger is sympathetic or not, and no amount of wheedling or pleading will make her change her mind. One summer, we had two painters decorating the flat. Minni ran for her life every time one appeared, but would keep the other company to the extent of climbing up the ladder with him when he painted the ceiling.

Curiosity is another of her personality traits. When I began shaving with an electric razor, Minni soon came to see what it was all about. Every day she watched me, at first from a distance but later approaching progressively nearer, until one day, plucking up full courage, she climbed right up to my cheek to inspect the machine from the closest possible angle. Once having done this, however, she lost interest.

On another occasion, I was taking photographs with a flash bulb. Miku almost jumped out of his skin as soon as I pointed the camera at him, but Minni stayed put, watching with interest, waiting for the flash. After one such flash, she jumped straight at my chest to smell the strange contraption. Thereafter, no more than a passing interest was shown.

It was from this curiosity that ingenuity stemmed. Minni watched how I took the lid off the small round seed box. Soon she was there helping me, pushing at the lid with her nose. She also noticed that there were other boxes in the kitchen, square ones this time, which also held food, and where the lids could be removed in the same way, with a little help from her teeth. If the kitchen cupboard was ever left open, Minni was soon inside, at the lids, trying her best to get them open. One I had to go out suddenly, leaving the kitchen cupboard open. On returning, I was greeted by small mountains of rice in the middle of the hall carpet. Minni had managed to open the container of rice, but the taste had not been to her liking, so she played with the contents instead.

Because of her tricks and a little black spot on the end of her nose, giving her an expression of mischief, I named her The Clown.

CHAPTER FOUR

Crow's Feather and other games

The first week of Minni's and Miku's life together they spent playing hide and seek, usually around an armchair. One would lift two furtive eyes above the back rest while the other peeped out from under the arm, and as soon as their glances met, they scrambled out to new positions; until again, having locked glances once more, new vantage points had to be sought.

Miku had always played alone until now. His favourite game had been played in the bedroom, and it was to wrestle with one of my old shirts. The ultimate aim was to drag the shirt up the curtain to his hideout. Walking into the bedroom unexpectedly, I would see what appeared to be a shirt dancing up and down in front of the curtain of its own accord. Though I knew of course that Miku was inside it, straining for all he was worth, I couldn't resist a quick pull, which soon resulted in a tug of war, lasting many minutes before Miku let go of the shirt. This was the only game Miku let me share with him.

But Minni and I have played many games together. When she first arrived, aged three months, I encouraged her to eat her cob nuts from my hand. Instead of eating them, however, I noticed that she stuffed them into her cheeks and carried them into the bedroom. This made me curious. Where did she take them? To a secret nest? So I followed her — and found out that I was being had. In a nice way, of course. She would wait for me to creep up stealthily and when she was sure I could see her, though pretending she was unobserved, she would go through the motions of planting a nut under the pillow. Her still puffed cheeks, however, gave her away. All the same, she was highly delighted when I, poor fool, lifted the pillow and found nothing. She would jump high into the air, do somersaults, and then would rush back under the pillow to emerge once more with a serious look that said, "But I left one this time. I did honestly!" She would then leap off the bed and scurry to the study, with me following of course, and begin the same game all over again, this time running at random over the bookshelves and planting a nut behind each book. And she really did have me! I couldn't keep up with her. Now she would leave one. Now she wouldn't. Until, in the end, all the nuts were finally deposited and I was none the wiser.

When I didn't follow her, she stopped by the door to wait for me. Asking me where I was. Why wasn't I playing any more? To be honest, I grew somewhat tired of this chasing, and when Minni grew bigger, that game was abandoned.

I must relate how I taught Minni to play cat-and-mouse, like a kitten. It all started when I discovered her great interest in chocolate. Quite by accident. We had just received a large box from guests, which we left on the radio-gram overnight. When I went into the room the next morning, the first thing I saw was a litter of silver paper all over the floor. As our guests are not usually in the habit of shedding wrappers on the carpet, I sat down in the arm-chair to await further events. I could see that the lid of the box had been pushed to one side — bearing Minni's fingerprints or should I say teeth marks. I didn't have to wait for long. Minni soon returned to the scene of the crime. Climbing up the back of the settee to the radiogram, with great confidence and greed she began to make her selection among the chocolates. She settled finally for one wrapped in silver paper once more, obviously her favourite — or was it? Having discarded the wrapper with a disdainful flick of a back leg, she tasted the chocolate, but, "Hmm!" it really wasn't quite what she wanted. Not knowing what else to do with it, she tucked it under one of the settee cushions. Then came back and tried another — perhaps this one contained a nut?

When I later lifted up the cushion, there were altogether five rejects hidden. After all, what do you do with a chocolate that you've already bitten into but don't want?

When Minni had departed, I pressed the lid tightly on the box. But I thought also, you wait! I'll put your greedy little habit to good use. I'll give you a run for your chocolates. I tied a thread to one of the silver-wrapped ones and then placed it on the carpet in the middle of the room.

Minni must have rubbed her eyes with disbelief at such good fortune. She came immediately, pouncing gleefully upon the chocolate . . . but I pulled the thread and poor Minni's paws were quite empty! For a moment she stopped, bewildered. What was this? Her chocolate alive? She became scared and jumped aside, her tail stiff and wagging with agitation. But curiosity was greater than the fear. Nose quivering, she came back to smell the sweet . . . again I pulled the thread. This time Minni followed it and the game had begun. Soon she had the chocolate captured, between her paws. Standing on hind legs, she started to remove the wrapper from her prize. This I guessed to be a good time to give the thread another jerk. However, I hadn't reckoned with Minni, who had no intention of letting it go again. Her claws were deeply embedded, and to have pulled harder would have meant breaking the thread as Minni was a little squirrel with a big grip. So it was time for a more sophisticated strategy. I let the thread go limp and fall in a slack coil at her feet and assumed an air of interest in other things. Minni loosened her grip at once and with what must have been a squirrel's sigh of impatience, set about the wrapper once more. But she had let go. And her paws were once more left empty.

This seemed to Minni a time for deeper

thinking. And she sat back to think. The chocolate was only a few inches away from her but she made no further move towards it — not directly anyway. She too was now pretending uninterest, while sidling towards it sideways — and that was how she discovered the thread. Up till now she had thought it was all due to the chocolate. She tested the thread gingerly with her paw, following its course, which led upwards. She even rose on tiptoe, still following the thread until she could stretch no further. So she continued to follow its course with her eyes — until they reached my hand. And from my hand — to my eyes. Ahah! That was all she wanted to know. "So you are the one behind this?" she seemed to say to me. Leaving the thread and chocolate where they were; she left me to seek other or new diversions in another room. It was a snub.

I had to admit, she had surprised me. She was pretty clever and that reminded me of an article I had read by an eminent zoologist, that in the event of the world being taken over by the animals, he himself would put the squirrel in second place instead of the ruling primates.

Minni loved to plant sunflower seeds into all our plant pots, irrespective of what they already held. She did this very neatly, scraping away a little top soil, put the seed down and then covering it up again. And to this day, I have never found out why she only planted sunflower seeds and nothing else. Not nuts, corn or any of her other food. She couldn't have known that only sunflower seeds actually take root while the others would have rotted anyway and their planting would have been of no use.

Miku soon learnt that Minni was hiding sunflower seeds in the plant pots and so he followed her, scraped them up and ate them. Before this, he had never been in the habit of digging soil, but now he wouldn't eat his sunflower seeds from anywhere else — there were plenty on his eating table. Perhaps the fact that most of the seeds were already germinating after a few days in the soil was giving them a better taste?

This digging in the plant pots was becoming quite a nuisance. There was always soil to be swept up and the plants themselves were starting to wilt and die as their roots were being gnawed at. I decided to buy Minni and Miku their own field to dig in, with high walls. A special box which I filled halfway with soil.

Minni claimed the new garden for herself at

once. She dug a few trenches, then destroyed them again the next instant, like a child with sandcastles. She rolled on her back and took soil baths. I don't remember having laughed so much in years as I did at her antics. And the more I laughed, the wilder Minni's antics became. Like a child in water, she began to splash me, throwing up soil with her paws while burrowing deeper and deeper with her nose until the soil on her nose made her look like a baby rhinoceros. So she would stop and look at me. If I was still laughing, she would splash me anew and then go into fresh con- vulsions of exhibitionism. This game some- times lasted a quarter of an hour, at the end of which she would jump on to the edge of the box, shake the soil from her fur, and tho- roughly wash herself. After each such bath, I noticed how Minni's fur shone. Miku, how- ever, didn't care for this kind of bathing and his fur remained dull in comparison.

Of course, Minni also continued planting her sunflower seeds into the new box, but as soon as Miku tried to dig them up, she drove him away. She couldn't abide to have Miku share the "sand box". In this way, the seeds remained put and as the soil was fairly rich, they soon started to germinate. Soon little

45

shoots began to push up that grew and grew. Minni was delighted. She wandered around her growing garden, smelling and testing the new shoots, without harming one of them. (In much the same way she often playfully bit my earlobes and fingers without causing pain.) My wife and I were very impressed with our little gardener. Fancy planting her own seeds so that she could reap her own harvest!

Alas, this didn't last long. A week later, a dirty Minni must have realized she hadn't had a bath and needed one badly.

I happened to witness the complete destruction. Young sunflowers and soil flew in all directions until the last shoot was razed and uprooted and the ploughed field was bare once more. Then Minni got down to her bath.

There's an old saying: You can't turn a goat into a gardener. I might now add, or a chipmunk either.

Unfortunately, Minni only used her "sand box" for games and baths until she was six months old. With new sophistication, she abandoned the childhood scene and moved into much more arty circles; in fact, into the soil base of our most prized *objet d'art* — a piece of mounted driftwood whose shape resembled that of a camel. Minni got right down under the hump, breaking all the ferns we had planted round it. She bathes there to this day. When no one can see her. (Perhaps she realizes we are not too pleased with her destroying our artistic plant arrangement?) She climbs out of there, dirty and muddy, to catch her breath after a frantic scuffle, and eventually sits there quite still until I try to coax her with a nut. Being a very polite chipmunk, though, she then quickly sets about sprucing herself prior to submitting to temptation.

And now we come to a really dangerous game which Minni and I play.

It started when one evening my wife came home from work with a crow's feather which she had picked up in the street.

"I brought this for Minni," she said.

Minni was just about to retire for the night but was still squatting on the edge of the armchair — thinking. Just in fun, I flicked the feather under her nose and against the armchair. I was unprepared for the way she reacted. In fact I was quite shocked. With a murderous flurry of claws and teeth, she pounced to kill it. This isn't Minni, I thought. This is a wild beast jumping on her prey. Chipmunks, it seems, may live on a diet of fruits and nuts but are apparently also partial to a little fowl meat

47

on their menu, if it happens to come to hand. It was quite obvious that the crow's feather didn't just rouse Minni to fun and games but to a primitive hunting instinct of considerable ferocity.

I learnt to move the feather so fast that it was always just out of her reach — and a wild chase would follow, across tables, chairs and other furniture, together with high jumps on the carpet.

It is not as if Minni doesn't know that it is my hand which holds the feather and guides its flight. She knows very well. She even climbs up my side sometimes, and along the arm that holds the feather. With lightning speed, I change hands and the "bird" disappears behind my back. This drives Minni to a pitch of excitement. And anger. She even gets carried away enough to attempt to bite the back of my hand. Luckily she usually realizes what she is doing before she actually bites, and besides, the back of my hand is too wide for a good grip. All the same, it is usually full of teethmarks and the game often ends with a few drops of blood.

During this game, Minni does not like me to laugh. I have laughed at her fruitless journeys along my arm and her vain leaps into the air — when suddenly she will jump on my chest and, ignoring the hand with the feather, start to attack my laughing mouth and in fact bite my lips. And on retiring back to the chair, she will sit there accusingly while I wipe the bloody lip.

"Yes. Serves you right. You shouldn't laugh at me."

Therefore, my face is now stone sober whenever the "crow feather" game is being played. I also try to keep it within sporting limits. As soon as I notice Minni's hunting instincts getting the upper hand, I end the game abruptly, substituting a nut for the feather. This calls a truce immediately, the hunt is forgotten, and she returns to her feeding table and selects a grape, sucking it empty as it were a glass of champagne.

"Cheers!" I toast her ruefully, licking my wounds.

CHAPTER FIVE

Of nuts and roses

Nuts, as everyone knows, are a vital part in the life of squirrels and chipmunks. The cracking of the nuts, however, constitutes a paragraph all to itself. My chipmunks Miku and his predecessor showed absolutely no dexterity in this direction. Miku has great skill in opening sunflower seeds but doesn't bother trying to open anything harder.

Before Minni arrived, I was under the impression that the chipmunk, being so much smaller than the ordinary tree squirrel, didn't have the strength of jaw necessary for cracking large nuts. When I first handed Minni a cob nut, out of curiosity, she accepted it with interest, twirled it round between her paws and then started nibbling the shell.

I had watched a programme on television about the ordinary tree squirrel, in which it had been explained that the art of cracking nuts is not a hereditary skill but one passed on by the mother. The young squirrel that stays in the nest close to its mother learns such things naturally. But if a baby squirrel is brought up in captivity, away from its parent, then it may have an instinctive interest in the nut it is offered but has to solve the mystery of getting inside it all by itself.

This process can take weeks, as I was able to witness — the trials and tribulations and eventual triumph of a young chipmunk trying to crack its first nut in my own home. Following Minni's initial curiosity about nuts, I encouraged her interest by offering them to her at frequent intervals. She chewed and bit at the shells but couldn't get inside them. She usually in fact carried them up to the windowsill with her, sitting there and gnawing away at them in vain — knack-knack-knack was the sound it made. I decided one day to time her efforts, but no sooner had she commenced than a pigeon disturbed her by flying close to the window. She grabbed her nut and fled, but as soon as the danger was past, she returned and persevered at her task, then the noise of a low-flying aircraft caused her to flee again, but she returned and carried on the old knack-knack-gnaw-gnaw routine. After eighteen minutes she gave up. The result of all the effort was a small hole in the shell which was mostly due to the outer layer having been worn wafer thin by the scraping of her sharp teeth.

My soft heart could not bear this much longer. With a nutcracker, I made small cracks in the nuts before giving them to Minni, planting a few whole ones in the pile. Minni opened the cracked nuts right away, putting the whole ones aside, though never carrying them into her nest. When all the cracked ones had been stored, she turned to the others. Then once more the flat sounded to the patient knack-knack-gnaw-gnaw of the hard nuts Minni still had to crack.

"She is learning," we noted, as the noises accompanied each and every one of our conversations. On examining the eventually rejected nuts, I could see that all kinds of techniques had been used on them but that the right one had not yet come to hand.

Suddenly, one morning, roughly a month after Minni's first tooth mark on her first nut, I noticed that not only did she peel all the cracked nuts but the whole ones as well, breaking those open just as easily.

At first I was a little sceptical, thinking that

perhaps I had inadvertently cracked all of them myself. I quickly got a fresh handful of nuts and offered them to Minni. She expertly cracked them open in half a minute.

Clever Minni! It was a time for jubilation! She had finally passed her test.

On the other hand, I was very interested to see wherein the trick lay, and found that Minni turned the nut so that the sharp end was to her mouth. She then bit at this sharp point, at the two sides where the veins meet. The end cracked easily and the whole shell broke in half as soon as her teeth got into the crack. I tried it for myself. It was easy. I too chewed at the sharp end; it was so much softer and the shell broke with comparative ease. Up till now I had always relied on the nutcrackers, and when they hadn't been to hand I used my teeth, almost breaking them, and always approaching the nuts on their rounded sides.

It had taken a chipmunk to teach me how to do it, something my own human brain hadn't been able to solve for me.

Minni had needed a month. A long time. But I expect it would not have taken half that time if I had not tried to be kind, giving her cracked nuts instead of whole ones!

There is one more thing I wish Minni would

teach me: how she knows, without opening the nut, that its kernel is bad and not fit for eating! I have offered her what appears to be a plump and perfect nut from the outside, which she has rejected. Thinking she is just being temperamental, I have mixed it with the others and offered it again and again, but no. Or she will finally take it, just to show me — cracking it carelessly and throwing it at my feet.

"Look for yourself."

And believe it or not, the kernel is either completely shrivelled up or covered in mould.

"You don't want it yourself, but you'll give it to a chipmunk," seems to be Minni's reaction when she sees me throwing it in the waste-paper basket.

I have tried to smell those rejects, even to weigh them in my hand, but have found no difference — but then of course I don't have a squirrel's nose! And for that reason, I suppose I'll never learn to differentiate between a good and a bad nut either!

Of course, Minni can't open every kind of nut. Especially the round varieties. She will roll them round in her paws, but as they don't have a sharp end they can't be opened. Neither can she manage almonds, Brazil nuts or wal-

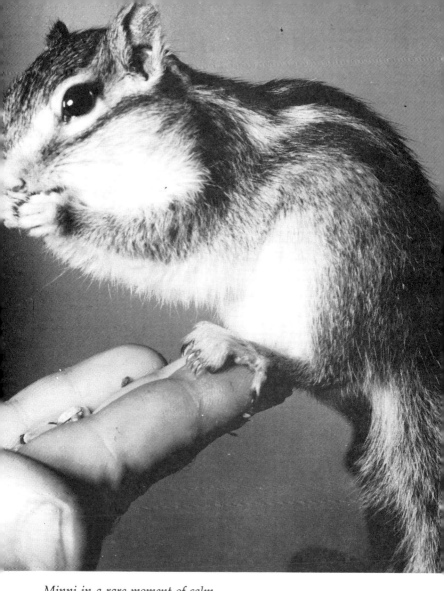

Minni in a rare moment of calm

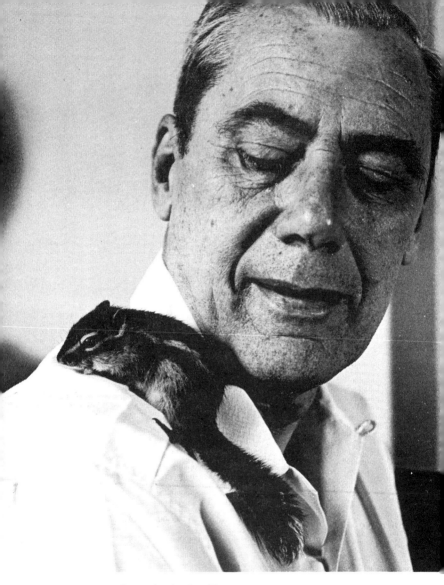

Minni on the author's shoulder

Minni enjoyed being fed her breakfast . . .

. . . but would just as happily fend for herself

Wherever there was food . . .

. . . Minni would not be far away

Miku takes more than one bite of the cherry

A delicate balancing act

Minni at home on the bookcase

Miku valiantly defends his home territory . . .

. . . but finds it hard-going against the more aggressive Minni

Minni always yearned to explore the outside world . . .

. . . and eventually escaped into the garden and the park beyond

nuts. Her strength just cannot compete with the toughness of their shells. I usually break those for her, and I have to be very careful; she participates in this operation so eagerly that I have to watch out that I don't catch her paws in the nutcracker. Once the nuts are half cracked, the honour of extracting the kernel is still hers, which she does right away while still on my knee. The walnuts I usually break in half, holding the halves open for her to scrape out the inside.

Oddly enough, Minni doesn't like this sort of "facilitation". If the nut should fall away from my fingers, she would rather hold it for herself. It only happened once, after the nut had fallen and she had already carried it away, that she returned unexpectedly and handed it back for me to hold for her. Taking into account a squirrel's or any wild animal's natural instincts this was a great trust indeed!

My wife and I have to laugh when we see Minni stuffing her cheeks with the oval, complicated kernel of the walnut. They are like long French loaves, with the ends for ever escaping from the basket. Minni pushes at the ends with both paws, her cheeks extended in weird shapes, till they are all inside.

The world's largest nut, however, the

coconut, is also a favourite of my two little chipmunks. I saw the nut into two halves (I drink the milk myself — this they don't want) and put the halves before them. Miku bites at the edges while the giant nut keeps slipping away. Minnie, however, soon found a sensible solution to this slipping. She jumped right inside it. It swayed a little, like a ship on a stormy sea, but she soon found her balance, and the coconut became as familiar as a tight-rope to a tightrope artist. (She sometimes hangs upside down from my finger like a bat. The back legs grip my finger easily with hardly any pressure. There she swings back and forth like a pendulum.)

The coconut meat is bitten away in a circle and carried to the nest. After about three-quarters of an hour, all that is left of Minni's coconut is a disc of the coconut which Minni then lifts up like a shield, as if to weigh it.

"Too heavy," is my comment from the side, as I watch all this with interest.

But in the next second she has already taken up this shell and fled towards the study with it, where her nest is. I follow to see how she manages.

First she tries the direct approach. She has to jump two feet from the ground to the bookshelf. She gathers herself for the leap but hasn't calculated on the weight of the coconut, and falls short.

The other way is via the curtains to the bookshelf, which avoids any jumping. She manages somehow to get up there but still has to carry on to the top of the cupboard — by jumping across a chasm of two and a half feet. She jumps but once more the weight of the shell pulls her short and both Minni and the coconut take a tumble of nearly seven feet. Minni of course falls like a cat without any harm done, but she loses the coconut piece in the process. Here I decide to be naughty. I pick up the coconut shell from right under her nose and carry it back into the living room, back to its original place. I have an idea. I want to take a photograph of Minni with her giant shell.

Before I have a chance to set up my camera, she is off again. This time to the bedroom, to the top of the wardrobe, seeing the last place was too hard. She has no difficulty climbing from the headboard of the bed to the dressing-table; from there to the wardrobe, however, there is still a space of two feet. She could try jumping once more. It is easy without the burden. But Minni already has experience behind her; she measures up the distance and decides not to risk it. Instead, she disappears under the bed with the piece of shell, only to come out again "empty mouthed", with a sly look on her face. I find the coconut hidden in my shoe. Once more I carry it back to the living room and put it in its original place. I have not yet got the picture I wanted. But I have to hurry now as Minni is already getting indignant and her chipmunk's pride is awakening; this must really be a special piece of coconut, she seems to be thinking. Otherwise the giant wouldn't take such an interest in it, or flash around with his camera every time I pick it up in my mouth . . .

No. This one really has to be hidden, come what may.

Three times more I manage to retrieve the coconut, pretending not to see where she hides it. Finally I let her put it on the bookshelf behind some books.

Minni is contented at last. The special piece of coconut no longer reappears in the living room.

What a chase! Minni slakes her thirst and proceeds to wash herself. Her chipmunk's honour has been saved.

So much for nuts!

Now to the roses. One hardly ever thinks of

chipmunks and flowers in the same context. In this case, however, roses in particular play a major part in Minni's life. Two friends came to visit us one Sunday, bringing a bunch of roses with them. My wife put them into a vase and placed them in the middle of the low coffee table. We were still in the middle of admiring them when suddenly, stealthily, Minni appeared on the scene. Rising to her haunches she lifted her paws to the flowers and, as we had all been, seemed enchanted by their smell. The guests were most surprised.

"What a lovely creature! And see how she loves roses!"

Hardly had the words been spoken — during which time Minni was not only smelling the roses but embracing them as well — than we heard a sharp chomp and the bud of one of the roses was in Minni's paws, the stem bitten through. At once she started to peel the rose bud, tossing the petals on the table around her until soon, standing among them, she made you think of the "flower people". She was after the heart itself, which she ate with keen enjoyment and would have reached for another . . . had we allowed her. As strangers were present, it wasn't difficult to chase her away. One of them had only to stand up and take a step towards her.

On another occasion my wife received a bunch of long-stemmed roses from our daughter for Mothers' Day. We put those well out of reach on top of the television set, hoping Minni would not notice them, but already she had skimmed up the curtains, which are unfortunately right behind the set, and before I could even reach the vase, she had already embraced the first one. Her back legs were still hanging on to the curtain. I pulled the vase away, but you should try getting something from a chipmunk that it doesn't intend to let go of. She let herself be stretched as long and thin as she could, embracing the rose with her eyes flashing protest.

"But this one is mine — I claimed it. Don't you understand? You can't take it from me."

In this case I did quash her natural instincts, and in the end my authority prevailed. A long, thin Minni had to let go and regain her balance on the curtain.

When those same roses started wilting. I stuck them in Minni's playbox. Minni didn't much care for wilted roses — but it was a case of that or nothing. Slowly she started to decapitate them as before, but the fun was all gone.

If one had a too thick stem which she couldn't bite, she got angry. Her tail swished back and forth and she started spitting tsukk-tsukk-tsukk, attacking the thick stem with her strong teeth flashing, all the while splashing me with soil from her back legs, while she stamped in rage. If she still couldn't bite the stem, after three minutes of temper, and fraying nerves, she spat for the last time and turned tail, fleeing in hot anger.

CHAPTER SIX

Minni's nine-day adventure

The time of our annual holiday arrived. We were going abroad. Our chipmunks were to stay in the flat with a friend of ours who was going to sleep there. He was quite willing to undertake the responsibility.

My wife was rather sceptical about this and suggested that the least we could do would be to put Minni and Miku into a cage and keep them there; but I, being soft-hearted as usual, could not bear to think of them being locked up. After all, they were used to freedom and secondly, their relationship was not, to my mind, agreeable enough to withstand close confinement in the same cage. They would be certain to fight, and who could say what we would come back to?

So they were left to run free after all. Unfortunately, our friend did not arrive until the day before our departure. So there was little we could do other than to instruct him what food to leave out and where it was kept.

The sash-cord windows were opened fractionally at the top, enough for adequate ventilation but not a large enough opening for an escape. We asked our friend to keep an eye open in this department and to be sure not to open the windows any further. This he said he would do.

Our vacation lasted two weeks. We returned on a Sunday evening, tanned and rested. The flat was in darkness, the chipmunks were obviously asleep. I was greatly tempted to go into the study just to say hullo and to tell Minni and Miku that their "giant" was back, but I didn't want to break our unwritten agreement that I would not disturb them in their nests. So I desisted, noting instead that the chipmunks' eating table was completely in order. So tidy, in fact, that my suspicions were roused. There were sunflower seeds, neatly piled, something Minni would never have left there, but would have carried to her nest before retiring. There was also an untouched heap of sweet corn. I went into the study to inspect the windows — it was dark outside — and my glance fell on a row of little black dots on the windowsill. On the *outside* windowsill. They were chipmunk droppings, and it looked as though whoever was out there, was there a long time. No need to panic, I reasoned with myself. After all, I often opened the windows myself and let Minni or Miku on to the windowsill, and they had never shown any inclination to go further. But, all the same . . . they never went to the toilet out there either, but sought more private places, like behind the curtains. (Chipmunk droppings, by the way, are completely odourless.) All this was most disturbing, suggesting that Minni (and it must have been Minni) had spent some considerable time squatting on the windowsill — but for what purposes?

I went to Minni's nest. It was empty. The friend, whom I woke up immediately, was quite indifferent about the whole thing. His answer was non-committal:

"Yeah! Sure I have kept an eye on them. But I have not seen Minni for days. Miku yes, but Minni disappeared a few days after you left, though I saw her again a few days later."

"But what about the open windows? What were the chipmunks doing outside?"

The days had been very hot, he explained. It had been necessary to open the windows for fresh air. He had not seen Minni on the windowsill, however. He was sure about that.

Knowing how fast Minni moved, this was not surprising. Besides, our friend was also very short-sighted. It was possible that Minni, on seeing a stranger approaching the window, had jumped to the nearest drainpipe. He had closed the window and Minni had been left outside, not risking another approach until perhaps hours later; hungry enough to face the strange man, whom she had obviously not trusted.

"As for the other chipmunk," our friend continued, "he has the oddest habits. He eats candles."

This was new to me! I went immediately into the living room to look at our candles — which were displayed as decoration — all six of them had indeed been almost chewed through at the base

Oh no! I had a dreadful thought and went back to the pile of sweet corn and sniffed it. A sour odour met my nostrils. Our chipmunks' sweet corn was frozen (this year's fresh corn was not yet ripe) and I had purchased it in an airtight plastic bag which was kept in the freezer compartment of the refrigerator. I had told our friend about this, but he had not taken me seriously, thinking the refrigerator compartments were all the same. He had put the bag on a lower shelf, where the contents had defrosted and started to rot. Fresh sweet corn is to the chipmunks what daily bread is to us. They are very sensitive also to the quality of freshness of their food, never eating leftovers the next day.

The following morning I went to see the caretaker of our building in the hope that he had heard something about Minni.

"Ah! So it was your squirrel that has excited the whole neighbourhood over these past two weeks!" he said.

"How so?"

"Let me see, today would be the ninth day since we first noticed her. Yes, it was a week ago last Saturday morning when a lady from next door came to me, terribly upset and frightened; saying she had just seen a huge rat enter the house from here. Rats in the building? I thought she must be mistaken so I didn't take much notice. But about an hour later, a gentleman came and told me a strange story of how he had got into his car and was just about to start the engine when a squirrel type of animal had crossed his front wheels and darted into our front hall. This was too much of a coincidence, so I started to keep my eyes open. Sure enough, I soon spied a small squirrel tucked into a crack in our garden wall. At the time I thought, this looks just like one of your chipmunks, but I didn't know you were on holiday and I thought that if you had lost one you would be sure to have come and told me — so I guessed this one must have escaped from the school down the road."

"When did you last see her?"

"Now let me see — day before yesterday. If I see it again, I'll let you know . . ."

At least she was alive the day before yesterday! This was some consolation to me. I still had a week's holiday left and ample time, I hoped, in which to find Minni safe and well.

It was now time to devote my energies to Miku. I hurried to the shop and bought a new packet of frozen sweet corn. Meanwhile, Miku had woken up and been at the candles again — and vomited all over the carpet. I guided him towards the fresh sweet corn and watched him attack it ravenously. Poor Miku must have been half starved. (Since then he has never touched candles!)

That morning I had a visitor and we were talking business but all the while I was restless and kept glancing out of the window.

67

Suddenly I saw her! Yes, it was Minni. Sitting outside on a low wall.

"Minni, Minni!" I called out of the window, and saw her rise to her hind legs. Although she must have recognised the voice, it came from too high up for her to place it, as our flat is on the third floor.

My visitor kept me a few more minutes, and by the time I was free to return to the same vantage point, I was only able to catch a glimpse of Minni's tail disappearing over the wall into a neighbour's garden.

As the gate to this garden was in another street altogether, I went to the house and rang the doorbell. A lady came to the door. I told her I had just seen my chipmunk going into her garden.

"So it's your chipmunk!" she exclaimed. "Thank goodness it has found its owner at last! My husband and I have been upset, worrying about what would happen to it when the cold weather comes."

Minni had become a regular visitor to this household, as they put a saucer of milk out for her every morning. She had also dined on their flower beds, eating, among other things, their roses. She had even crept into the house on one occasion and slept behind the piano, leaving her visiting cards all over the shining parquet floor. The lady was, nevertheless, very understanding and as Minni had already drunk her milk for that day, she suggested I might spend the afternoon in her house and garden. She herself was going out.

This was a very kind offer but I decided it best to return later, when she herself was at home. I left a handful of sweet corn, some nuts and sunflower seeds beside the saucer.

I saw Minni again that afternoon, from my window. She was coming out of the garden where I had left the food. A flock of birds were hovering round her and I remembered the lady telling me that Minni seemed to be attracting birds (perhaps in the hope of taking what morsels she might leave behind).

I called again, and again she stood erect to listen, this time even trying to get nearer to the voice by climbing a tree in the garden. Her vision was, however, obscured by the leaves and she didn't see me. Later I saw her going into the neighbour's tool shed.

Seeing that the lady was now home, I decided on a plan of action. I filled the cage with Minni's favourite food and took it into the garden, placing it next to the saucer of milk. I left the door open, tying a long piece of string to it, so it could be closed from a distance. The lady had been busy telling me how much more timid Minni was each day — hence a very long piece of string was needed. The end of it we tied to the bathroom window hook, and the lady very kindly agreed to watch the cage that evening so that as soon as she saw Minni go inside to feed, she would pull the door closed. I did not think Minni would return that evening, and promised to be back myself the next morning, as early as possible.

Minni did however come back, probably remembering the unexpected snack earlier. At seven o'clock, glancing out of the window, I saw the neighbour creeping towards the wall where the cage was. And the neighbour's wife pulled the string.

Got her!

I ran next door and there Minni was, sitting still. She let me come near without any fuss until I lifted the cage, which upset her, but only for a minute. The couple were now able to admire Minni at length. I thanked them for their kindness and returned home.

Placing the cage on my desk I opened the door. Minni emerged immediately but was somewhat confused, running straight to the window and hitting herself against the glass.

Recovering and realizing where she was, home at last, she made her way across the bookshelves to her nest, where she slept till lunchtime next day.

Her nine-day adventure was over.

CHAPTER SEVEN

An incredible story

All to whom I have told the following story have seemed slightly sceptical, and yet it is quite true. On the other hand, had someone else told me the story, I might have found it just as unbelievable — who knows?

It was a week after Minni's "nine-day adventure". Our relationship was as close as before, though the morning after had been a little strained. That lunchtime, however, had found her in the usual place, eating from my hand and even permitting me to stroke her.

Minni showed no interest in open windows, but at the end of the week I decided to test her. I opened a window and stood by to see what she would do. At first — nothing. She jumped on the windowsill, looked round, and as soon as I called her, came back into the room.

All the same, I began to make sure that when we left the flat, all windows were securely closed. In moments of boredom she might suddenly remember her nine days of freedom, and having once had a glimpse of the outside world, might want to explore further — seeing it had not been too bad!

Out walking one afternoon, I suddenly remembered an open window in the living room but hardened my heart and didn't turn back, but continued my walk.

When I got home Minni was waiting for me by the desk, sitting on the back of my chair washing her face.

Our daily routine returned to normal, with my usual calls at the printers. On leaving the flat I still closed the windows, although not the spare bedroom's window as the door was closed.

Returning from one such trip, I could not find Minni anywhere. Only Miku, climbing up and down the curtains on the "forbidden side" of the room. This immediately made me suspicious, though perhaps Minni was taking a nap? I waited an hour and then got restless! No sign of Minni anywhere in the flat! I also discovered that the spare bedroom's door was ajar; a wind had risen and opened the door, which had a faulty catch. I was greeted not only by the billowing curtains but by a tell-tale sign that suggested Minni had been there: a discarded aster which she had carried in from the living room and with which she had played "loves me, loves me not" on the divan.

But what then?

I looked out of the window — my eye caught the drainpipe directly next to it. My anxious gaze then travelled down to the wall and into the neighbouring gardens. There was no sign of life.

"Minni, Minni!" I called outside and inside, as I was not sure she had actually left the flat. The wardrobe door was also ajar; she might have gone in there as she had done once before, sleeping all day until evening. In such cases she never responded to my calling either.

The evening approached but still no Minni; she did not even appear for her meal. Her nest was also empty. There was no alternative but to face the fact that she had escaped once more, out of the spare bedroom's open window.

In the evening I went outside, prowling round the back yard and searching for a certain crack in the wall, the one which the caretaker had told me Minni had used as a store room during her first escape. A few stones were loose and the hole quite deep. I called and called, rattling the box of nuts, but to no avail.

The next morning I had to spend at the printers, so did not get a chance to resume my search until after lunch. There was still no sign of Minni, so I went to the caretaker to tell him I had lost a chipmunk once more.

"I am afraid I can't help you this time," he said. Together we searched through the whole of the back yard, in every nook and cranny. I even called on the lady next door to look into her garden. But no; this time Minni seemed lost for good.

I returned to the apartment — to hear a funny noise coming from the chest of drawers under the chipmunks' dining table. A sort of "chip-chip-chip". Minni is in a drawer, caught up in a ball of wool or string or something, was my first thought. I began to empty out the drawers one by one, frantically searching through their contents. But standing facing a mound of odds and ends, and no Minni, I realized I had been deceived by the chirping of birds outside. I was slowly replacing the contents of the drawers when the doorbell rang.

The caretaker stood there. Very grave and serious.

"I'm afraid I have some bad news about your chipmunk," he said.

"Why, what has happened?"

"This time, it seems, she did not climb down as before but up to the next floor."

Of course, I thought. Minni already knew what was down there and decided to explore higher. It was all in keeping with her curiosity. But what had happened? Had she fallen down?

"The couple living above you have only just moved in. They have two children. They didn't know you kept chipmunks," said the caretaker apologetically.

"What did they do?" I asked.

"Well, on hearing the child screaming they went into the bedroom to find a squirrel sitting on the edge of the cot. Their immediate reaction was that this was a baby squirrel which must have fallen down through the chimney. There were no open windows in the room. So they telephoned the R.S.P.C.A., telling them the story and asking them what to do. The R.S.P.C.A. suggested they try to catch the squirrel and take it to Holland Park. This they did; they caught it last night and took it to the park, where they let it go."

Holland Park! Now I began to realize why the caretaker was so serious. Holland Park was not only a good distance away but was one of the "wildest" parks in London. It is about a square mile in size, comprising gardens, an open-air theatre, Lord Holland's palace, an open-air concert platform, a peacock garden, a children's playground and a youth hostel among other things. The trees were huge, full of birds; even the night owls. But not one squirrel. There were no fir or pine trees or other nut-bearing trees. It was summer. Minni could live off the flowers as before, or perhaps find some acorns. But her life was now in peril in any case, even if the night owls had not eaten her already. To look for her there was as hopeless as searching for a needle in a haystack.

I tried all the same.

Holland Park was a favourite strolling place of mine. I often walked there and fed the birds. I was already friendly with a pair of Chinese geese, who always came to greet me with loud cries. I went to find the Head Warden, to tell him my story.

It was still early afternoon. The park was quiet. The giant trees rustled gently in the breeze and it became all too clear to me that this search was quite hopeless. How could I ever find a little chipmunk in all this? A gardener pointed the way to the Head Warden's office. I was lucky, he was there in person. No, he said, they had not seen a chipmunk nor had anyone reported seeing one. I even had a photograph of Minni with me which I left with him to show to the gardeners and other wardens when they came in for their tea break. He promised to do all he could and to ring me as soon as Minni had been sighted.

I walked through the park, along the quiet paths. I asked two ladies feeding the birds whether they had by chance seen a chipmunk — but they had not, though they had been there a couple of hours. They too promised to keep a wary eye.

I went home for tea.

The doorbell rang. It was the lady from upstairs; she had just learnt from her husband by telephone the exact spot in the park where he had released Minni. Apparently it was at the same entrance which I myself had used earlier. On his opening the box, Minni had immediately jumped out and scrambled up the nearest tree, disappearing among the branches.

The lady was most upset and embarrassed, saying, "I have been praying that you will find your squirrel."

In this case, however, I doubted whether even prayer would help. I went back into the park, this time armed with a box and plenty of food. I stopped under a large tree next to the gate, the one Minni was supposed to have climbed, and called, without much hope:

"Minni, Minni!"

Only a bird glanced down in astonishment.

Two brown uniforms were approaching. One turned out to be the Head Warden with whom I had spoken earlier. Yes, he had informed all the other wardens but I must realize how difficult it was to spot an animal of that size. As if I didn't know!

"I suggest you climb the fence, into the thicket," he suggested.

Most of the sanctuary areas were fenced off. Anyone seen climbing them was prosecuted. Having now got permission, I climbed over willingly. I continued slowly through the long grass, calling "Minni, Minni" as before. Last year's leaves rustled underfoot as I got deeper and deeper into the wood. Once I thought I saw Minni, jumping in the tall grass, but it turned out to be a bird, eyeing me with curiosity. Quite tame. There were birds everywhere: blackbirds, wrens, robins — but no Minni. Some tree-trunks were hollow, others had knot holes in them, inhabited by moles, hedgehogs and shy field mice. Perhaps Minni too had found a hollow tree-trunk? I sprinkled a few sunflower seeds here and there. My journey took me to a large pile of deck chairs, used for the summer concerts; suppose she was hiding among them? I placed the box, with the nuts and straw, next to the chairs. To the right was a rose garden. I searched through the roses, but again no. At the other side was another coppice as thick as the first.

I turned back. Who was to assure me that Minni had actually stayed in the trees or even in the park? There was a school with extensive grounds right next to the park, which was empty for the summer holidays. Next to that the houses began, each one with a back garden. Who was to say that Minni had not remembered her earlier adventure and sought the more familiar surroundings of a private garden? I approached the path once more, meeting the two ladies I had encountered earlier. It seemed that they too had hurried home, had their tea, and returned immediately, to help me find Minni. They also told me that their search had taken in all the wildest and densest parts of the park. They had found nothing.

My hunt continued — I was not yet willing to admit defeat — to the other side of the enclosure. Here too I looked into tree-trunks and hollows, depositing a few seeds and nuts at each, even in front of a tool shed, just in case.

Sitting down on a tree stump to regain my breath, I saw that the afternoon sun was already beginning to sink behind the trees. Except for the chirping of an occasional bird, everything was still. Peaceful. Was not this where Minni really belonged? Wild and free. Perhaps even at this moment she was jumping about somewhere close by, happier than she had ever been with me, rising on to her hind legs to chirp in chorus with the birds, in her true environment at last. Should I not leave her here? My mind was in turmoil.

With mixed feelings, I rose to my feet again. It was time to go home for dinner.

As I neared the fence, to climb over it back to the path, I noticed a tall grey-haired gentleman walking along. Arm outstretched, he moved along the fence which ran close to the bushes and shrubbery on the other side. There was a great rustling and thrashing in the bushes, and as he brushed past, scores of birds flew out to meet his outstretched hand, to eat from it.

Birds of every colour that I could imagine and more, fluttering and hovering around him; and he kept replenishing the handful by dipping into his pocket again and again.

He might help me, I thought. I quickened my pace and caught up with him. I was still on the other side of the fence.

I told him my story.

"Why don't you look into the peacock enclosure," he advised me. "If the chipmunk is used to being fed by humans, it is bound to go there as people visit that place most frequently to feed the birds." I noticed he spoke with an American accent.

"But that's at the other side of the park — she was released on this side," I told him doubtfully.

"Doesn't matter. Instinct would have guided her there. All the hedgehogs gather there, especially early in the morning and late at night, searching for leftovers."

Having given me this information, he continued his walk, arm outstretched, with a swarm of birds around him.

Deep in thought, I went home and had dinner. Afterwards I decided to go back to the park. I took with me a small box of sunflower seeds. It was already half past seven. At home

Minni usually went to sleep before seven, but in these unfamiliar surroundings perhaps anything was possible . . . or perhaps it would be more realistic to admit defeat? However, I went.

On entering the park, the first thing I did was to see whether the food in the box and tree hollows had been touched. It had not. Not even by the birds. I walked through the rose garden to the concert platform and the empty lawn before it, which was filled by the chairs during performances. Then across the lawn to the peacock enclosure — a large area of wilderness enclosed by a low fence. Here the peacocks had their nests. Also my friends the Chinese geese, who were already sleeping. A few male peacocks still strolled about, their tails neatly folded.

Here and there a few exotic hens and roosters pecked at the ground in competition with the usual swarm of pigeons and sparrows.

It was a beautiful evening and had invited quite a few late visitors, who paused in their walking to feed the birds. I stood beside them and directed my gaze to the denser parts of the enclosure, to the edge of the wood where the last rays of the sun still glinted. My eye caught, what looked like Minni balancing on her hind legs. Incredible! I kept my glance fixed on whatever it was until the very immobility of it made me doubt if it was anything more than a fallen branch. But this surge of hope made me scan the enclosure with renewed vigour, delaying departure as long as I could. I must have been there a considerable time before finally starting to walk back, past another similar enclosure where scores of pigeons and sparrows had congregated in hopes of being fed.

I hardly gave them a glance when suddenly . . . in the long grass, in a patch of light, something jumped. It could not have been a bird. It was Minni! She was approaching directly towards where I stood, hurrying as fast as her jumps would carry her.

"Minni, Minni!" I called, as soon as I thought she was within hearing distance. She stopped, rose to her two feet, looked right at me and then continued her jumps.

I began to back away, to get her to come to the path where it would be easier to catch her. Standing in the middle of the path, I took some sunflower seeds from the box and, calling her by name, offered them to her.

I need not have bothered.

She stopped once more by the fence, letting

a person pass by, then came on immediately through a flock of pigeons, dodging their sharp beaks, to the path and straight to my knee. A few lightning seconds later she was on my hand, greedily attacking the sunflower seeds, stowing them into her cheek baskets. Despite the fact that she was probably very hungry, she dared not yet eat them; the surroundings were strange and people kept passing.

Meanwhile I took out my handkerchief and with a quick movement covered the eagerly dining Minni with it. This was a stranger than ever experience for her but she did not seem to mind, and kept on filling her "baskets". I closed my hand around her gently, knowing I had to keep a fairly firm grip. It was too much to expect her to ride home with me perched freely on my hand. A dog's barking, a stranger's curiosity, my own movements — anything might have caused her to panic. I gripped her firmly, underneath the handkerchief, but of course my grip was not hard enough and she jumped out, first on to my shoulder but then trustingly back to my hand.

People walking by had stopped, staring silently at this strange spectacle. I had to come to a decision. If she eluded my grip a second time, she might panic and run back into the wood. I covered her with the handkerchief once more and applied a strong vice-like hold to her middle. I had her at last — but in panic. She struggled for all she was worth, pushing the handkerchief away. Suddenly her head appeared between my thumb and forefinger. She could have bitten me hard, as the first chipmunk had done when I took him to the veterinary surgeon. To my surprise, however, she did not.

I closed my hand around her, leaving her head and forepaws free with the thumb and forefinger tightly around her neck. This hold was most uncomfortable, both for me and for Minni. The back part of her body was swinging loose. I placed my other hand beneath her, but this gave her a boost and she strained to escape all the more vigorously. My grip around her neck had to be tightened. I know that chipmunks are so constructed that once the head is through, the rest of the body follows easily. After all, she was still a wild creature, willing to fight for her life with all her strength and to the last breath if necessary. The hold on her caused all the sunflower seed to be spilled out of the "baskets".

"This is my pet chipmunk," I felt it necessary to explain to the crowd that had gathered, and asked one gentleman whether he would be so kind as to pick up my spectacles from where they had fallen and restore them to my top pocket. The stranger obliged, but placed the reading glasses on my nose. I did not stay to explain further. To get home with Minni was all I had to do now.

I remembered the box with the straw, by the deck chairs. I would pass near the spot but to get to it would necessitate climbing the fence. Not with Minni. Perhaps I could stop a taxi once in the street?

I was in such a state of excitement that my glasses misted over and I could barely see where I was going. I asked a passer-by to remove my glasses and put them in my pocket. This he did.

Minni had become quite calm by this time, staring at me with her huge dark eyes. I was already outside the gate and on the paved path that led to the main road when I noticed that her eyes were beginning to close. I loosened my grip immediately but her eyes remained the same: half closed and glazed. I opened the whole of my palm and Minni lay on my hand limp and lifeless.

Had I found her, let her come to me trustingly, only to strangle her to death? Minni!

Minni! I coaxed her, breathing into her nostrils. Her little heart was still beating . . . and slowly she recovered. Thank God! She had only fainted. I realized that as soon as she felt well again she would start to struggle. I tied the handkerchief around her, comfortably. Not too soon!

It was easier to hold her now, though not much more so. I scanned the road anxiously for a taxi, but this was a quiet residential area with little traffic. I tried to stop a private car but my signals went unnoticed.

Somehow, with Minni struggling in my hands, her tail hanging out of the bundle, I made my way homewards, taking all the short cuts. There were few people about. In about ten minutes I was in the lobby ascending in the lift . . . then at last at my door. Minni made a final effort, having been frightened by the banging of the lift gate. I pressed my elbow to the doorbell and my wife let us in.

I placed Minni gently on the hall carpet. She did not stop to look back but fled to my study and into her nest. All remained quiet.

Looking back, it seemed quite incredible; I had gone to search for a needle in a haystack and found it — a tiny chipmunk in one of London's wildest parks.

Minni's latest adventure, however, had its
aftermath. The next morning, at the usual time,
there was no sign of her. Much later there came
the sound of rustling and nibbling from the
nest. It was most unusual, and I began to
worry. Perhaps I had injured her after all?

A couple of hours later she emerged, as light-
footed as ever, across the bookshelves, but
stopped above my desk to look down at me
with suspicion, her tail wound like a scarf
around her neck. After that she jumped past
me and went into the other room. She would
not eat from my hand that day.

Not until the next day, at lunchtime, did she
finally return to her old place on the back of
my chair. I called her name and she was round
my neck, as trusting and loving as ever.

But from now on, only with me. She had
ceased to trust people and when visitors came,
she no longer behaved normally as before,
when sometimes she would show off by climbing to my shoulder to crack a nut. Now, just
the ringing of the doorbell was enough to send
her running to her nest.

The telephone, right at her ear, made a much
louder noise than the doorbell, but this she
didn't even seem to notice, knowing perhaps
that this way no strangers could actually enter

the house. The doorbell, however, always heralded visitors — who might try to catch her, put her in a box and carry her away: a fear she had not known before. Even if someone came to my study, to watch Minni eating from my hand, she abandoned her meal and fled as soon as the stranger reached the door. As time passed, she got a little bolder, looking up at me for guidance . . . shall I run or shall I stay? If I persuaded her to stay, she would. But she would not even accept a nut from my wife's hand during that time of suspicion, which lasted over a month. After that, a few selected visitors were allowed to view her from a distance.

It was six weeks before Minni made her first reappearance in the living room, in the presence of guests. Another two months later, during a game of bridge, she climbed on to the table and accepted her first nut from a strange hand.

It seems that wild animals are very sensitive and remember a bad experience for a long time — that much too I have learned, during my life with chipmunks.

A month after the Holland Park incident, the phone rang. My wife answered. The caller was one of the two ladies I had met in the park while searching for Minni. She apologized for disturbing us, but her friend, who had been a visitor to England at the time, had written to her recently, expressing concern and wondering whether I had found my chipmunk.

I was able to tell them I had. Minni was safe and well — at home. It seems that I was not the only one concerned with her welfare!

CHAPTER EIGHT

The most bitter pills

Squirrels and chipmunks hibernate during most of the long winter and have to forage and store food during the rest of the year, so there will be enough to eat until next spring, when they awaken with nature.

Minni's preparations had begun early. Being a female, she had strong instincts in this direction, collecting food not only for herself but also for the possibility of young ones too. In contrast to her summer hoarding which she had deposited at random throughout the flat, she now concentrated her collecting on an adequate supply of food in her nest. This, as can be imagined, was no easy journey with bulging cheeks, across "miles" of carpet and "mountain ranges" of furniture.

During the autumn I secretly burgled the contents of her nest on a few occasions. Her complete hoard weighed nearly six pounds, of which I left about two pounds in her nest.

Minni did not seem to mind these raids too much. The first time, when she went to her nest to find it almost empty, she stayed there a little longer than usual, scrambling around. I had made sure that I had not disturbed her sleeping quarters, however. Having inspected her stores, she came out with new vigour to accept food again from me. So the same nuts made numerous round journeys — and Minni had the joy of collecting and I the joy of giving.

Miku, however, was a carefree batchelor . . . no food worries for him. He did not even appear to bother, probably surmising that the "giant" was there to provide and see to it that his table was always full.

In September with cooler days Minni began to look at the papers on my desk with new interest, testing a few sheets of my manuscript for pliability. I knew from experience with Miku the First what she was looking for and quickly placed a tissue handkerchief in front of her. Thankfully she accepted it, rolling it deftly into a ball — and taking it to her nest. Soon she was back for more . . .

Sometimes the "snowball" which she carried across the bookshelves fell apart or got caught in her feet. She would fall down together with the Kleenex, but there was no need to worry that she might hurt herself. She always landed on her feet, rolled the Kleenex back into a ball and resumed her journey. If the "snowball" kept on falling apart, she would become really angry, tearing it in half and making two trips.

I had to admire her practical instincts. She never made a trip unless she was absolutely loaded not only with the Kleenex but with her baskets full as well. No extra travelling for her. At night I could hear her tearing at the paper, chewing it into small pieces to line her nest.

At the first sign of frost, Minni came back to my desk for some more paper, this time even trying carbon paper. It rolled up nicely, but when she discovered the black smears on her paws, like a good housewife she abandoned it and started looking for white sheets. I gave her Kleenex once more, until she discovered the box for herself, on my wife's dressing table. Two days later it was empty. So! Time to face the long cold winter at last.

Miku, in his predecessor's ready-made woollen nest of a travelling rug, only made a few minor interior changes — gathering most of the material into one corner, into a tight wad.

However, it could not be said that only

female chipmunks are good nest-builders. Miku the First, a male, was a craftsman. The only problem was that in searching for suitable materials, he chewed at the curtains and blankets — a habit I soon discouraged. His nest was crocheted in fine strands of cloth and straw — a real masterpiece! This was situated in the cage, where the door was always kept open. Before retiring he always shook up his "pillows" and arranged his crocheted blanket into a tent shape, to circulate the air while he slept.

I am very sorry to say that on the misguided advice of an animal-lover friend of mine, I was led to destroy this masterpiece of architecture. As I explained earlier in the book, Miku the First was not healthy on arrival; he had a habit of chewing his hind leg. The animal "expert" of mine, who had some experience with guinea pigs, suggested that the cause was a dusty nest. It would be necessary to change the materials in the lining. With heavy heart I did as he told me, throwing out Miku's beautiful nest. I laid new materials in its place. Miku, however, was greatly upset and abandoned the cage altogether, carrying the materials to another site, in the seam of a curtain. It was a long time before he forgave me for this injustice — which it was.

In the wild, the chipmunk's nest is a many-roomed house, which he builds underground, with the entrance in the seclusion of a thicket or under a fallen log. First he burrows straight down into the ground for one or two feet, scattering the earth all round with his paws. When he can no longer dispose of the earth like this, he collects it in his cheeks and carries it to the edge of the hole. As soon as the chipmunk has burrowed down deep enough, he changes direction and digs sideways and upwards for another yard or two, but still deep enough in the earth for it not to freeze up in winter. When he meets an obstacle he digs round it.

Once the little architect has decided that the passages are long enough, he starts to widen one of them into a storeroom, about six inches in diameter. The storeroom ready, he starts work on another passage and another storeroom. Still digging slightly upwards towards the surface he begins work on the biggest room, the living room, then the bedroom and next to that another handy storeroom.

The chipmunk spends the winter in his small home, waking up from time to time to fetch titbits from the store room, nibbling at them, and then going back to sleep, curled up like a cat, with his tail as a blanket, until spring arrives.

Why does the chipmunk build his living quarters on a higher level than the passages? He knows that if his house is flooded the water will run down and collect in the cellar and not in his bedroom.

However, once the living quarters are built, it does not mean that the home is finished. From the bedroom he starts a complicated passage upwards until he reaches the surface, but instead of leaving the earth near his back entrance, he carries it laboriously to the place where he first started digging. At first sight all this labour might not seem very clever, but it has a definite purpose. The little landlord now almost always uses his complicated back door, and to leave a mound of earth near it would be as good as having a name plate and would tell unwanted visitors that someone was there. So he builds this back entrance in order that he can slip out unseen if an unwanted visitor arrives.

Once the house is ready, he furnishes it with dried leaves and grasses so that he has a soft bed to sleep on. And now the chipmunk can think of having a family. The young male and female chipmunks build separate nests, and then the female sings her love call to attract the attention of the male. The courting is quite elaborate, with the males fighting over the females, until a couple get together.

Now they move into the female's nest and live together until the young are born. Then the female chases the male away. "You know nothing of looking after children — you're just a nuisance." The male moves out, sometimes under protest, and goes back to his bachelor flat.

When the young have grown up and started to build their own nests, the female calls her mate back and they live together until more young are born. This pattern continues for seven or eight years until they reach old age. During the summer the female can raise three litters of four or five young ones, so the male becomes quite used to moving house.

Reverting to Miku and Minni . . . what about their imminent hibernation?

The flat is warm and because of this, the chipmunks do not actually hibernate as they would in the wild. All the same, they do respond to winter. The colder it becomes outside, the later Minni is in getting up in the morning, and the earlier she retires at night. In mid-winter, Minni rises at 11 a.m., eats peas and grapes, drinks some milk, accepts a few titbits from my hand, and goes back to bed for another hour and a half. She then makes one more appearance for a short snack. By 3 p.m. she is usually fast asleep. If the day is particularly grey and overcast, she does not return at all in the afternoon, but if the sun is shining, I have known her to stay up three to five hours, full of fun and mischief and even willing to partake in a game of "crow's feather".

Minni had another misadventure, just before Christmas. This one could have proved fatal.

It was a very cold day, dark and dull. Suddenly I heard Minni squeaking in the living room. This was almost unprecedented. I rushed to see what was going on, and to my surprise I found her running up and down the curtains, screaming all the time. I went to her and when at last she paused for breath she was shivering and trembling all over. After running hard she usually trembled, from excitement, but this time it was more as though she suffered an ague. I offered her a nut. She came to me but in a queer way, staggering sideways in another direction. She stopped, pulled herself together and went back to the starting point, to start again more directly. She cracked the nut as usual, when suddenly, unexpectedly, she attacked my hand, biting as hard as she could. She had never bitten me in anger before, not even during the game of "crow's feather". Having drawn blood, she scurried away to her nest.

There was only one solution. Minni had gone mad. But why and how?

It did not take me long to find out. At lunchtime, going to the kitchen for my blood-pressure pills (I always put out six yellow tablets first thing in the morning, to take two three times a day), I saw that four pills were gone. Knowing I have chipmunks in the house, I usually place the day's supply under an inverted tumbler. I knew I had taken two that morning . . . but then I remembered that when counting out the pills, I noticed there were enough for only one more day, which meant a visit to the doctor that evening. It was clear that I must have forgotten to replace the tumbler. So Minni had eaten them. She might have eaten all four or perhaps just one, taking the rest with her.

I hurried to look inside her nest. Minni was up and awake, sitting on the edge of the wardrobe watching me. She was apparently now quite normal. There was no need to inspect the nest, I decided. Whatever had happened, she was fully recovered.

That evening, at the doctor's, I told him that my chipmunk had eaten four of my blood-pressure pills. It was no reassurance to hear that four such pills, consumed by a small animal like a chipmunk, could prove fatal. The next day, when Minni was again in the living room, I went back to the nest to see whether there was any trace of them in her food stores. I found no tablets but instead, signs of fresh vomit. Something which was out of place in Minni's fastidious housekeeping. Whatever poison she had swallowed, her body had apparently rejected straight away!

This accident too had its echoes in the chipmunk mind. For a whole month she stayed out of the kitchen, the usual scene of her daily activities. If Minni could have talked, she would probably have said:

"They were the most bitter pills!"

In the winter Miku, on the other hand, sleeps until lunchtime and then appears and sometimes meets Minni at the dining table. Apparently a truce was made for the winter. I was only hoping it would last through spring and summer too.

Then suddenly spring arrived. To Minni it comes at the end of February. On one warm sunny morning she started singing her chirpy song, sitting on the window ledge from morning to night.

I shooed Miku out of his nest, inviting him to go and listen to how invitingly Minni sang! But to no avail. He did not show any interest at all.

I cannot claim to understand the chipmunk's language, but after long observation I have come to the conclusion that, just as my four-striped and five-striped chipmunks have not mated, so also there is a definite language barrier between their species. I can tell when they chirp for joy or squeak out of fear, when the voice is cooing tenderly and calling, and when it is angry and abrupt. But rather than attempt to add my own chirping to the confusion, I speak to them in the human tongue, which they seem to understand quite well. As when for instance I gave Miku the First a long lecture about chewing curtains, raising my voice and repeating his name over and over again every time he started. He understood well enough, and thereafter reverted to sly tricks, like chewing while I was on the telephone. I had to cover the mouthpiece and bark out his name before he finally desisted.

But returning to spring. At the beginning of March I noticed that Minni busied herself nearly the whole day in her nest, whence came sounds of great activity. Sometimes she appeared on the edge of the wardrobe with a piece of old Kleenex in her mouth and then disappeared with it behind the suitcases stored on top of the wardrobe. Was she moving her quarters to the other end of the wardrobe?

This activity went on for a whole week, and she still seemed to be nesting in her cardboard box; then one day when she was in the living room, I went up to inspect her nest. And what I found was that Minni had carried out a thorough spring cleaning! All the nuts, sunflower seeds and corn which she had collected during the autumn had disappeared. She had carried everything to the back ledge of the wardrobe and dropped her stores into the "crevasse". The same had happened to all the Kleenexes, which had formed her eiderdown. Spring was here: my swimming trunks were enough to keep her warm now. And the floor of her nest was as clean as if she had used a broom.

I started my story about Miku and Minni with events which occurred in the spring and autumn. Now the cycle is complete.

The new spring brought with it new adventures. Minni escaped again through an open window, and when she stayed nicely round the building I thought I would leave her in freedom over the weekend before trying to catch her. She spent the night in the hole in the wall familiar to her from the previous year. But on the second day I saw her crossing the street and a big dog started chasing her away from the house; but Minni did an about-turn right in front of the dog's nose and scurried in through the railed gate of our back yard, where he could not follow.

An hour later I saw a fat ginger cat sitting on the wall watching Minni's movements. I called out warningly, "Minni, Minni!" but she would jump coquettishly in front of the cat before running to another direction. Fortunately the cat was too well fed and lazy to start a chase.

But my nerves couldn't stand any more so I decided to entice her back to our flat with a few of her favourite morsels.

On reading through this story and pronouncing it finally finished, I decided to show it to Minni — who had already come to my desk to read over my shoulder.

"All right?" I asked her. "Seeing you have been present during most the writing of it, you might as well read it and sign your approval."

I drew some lines on the bottom of the paper, placed an ink pad in front of Minni's feet and asked her to be so kind as to give me her signature. The first time, she jumped on the pad but missed the paper, apparently on purpose.

Oh dear! Quickly I realized my mistake. I added the following last paragraph . . . after which she jumped back on to the pad and willingly signed it as correct:

"I, Minni, hereby certify (with my hind legs) the foregoing story, written by my friend the big giant, about myself and Miku, to be substantially correct (despite the fact that he is known mainly as a fiction writer). I also harbour one misgiving about his ability to read our minds. That part I will not sign to. Nobody knows what is in a chipmunk's mind except another chipmunk."

ASHFORD
Animal Classics

Charles: The Story of a Friendship
Michael Joseph

Lung Chung: The Diplomacy of a Pekingese
Margaret Ironside

Six Came Flying
Marquis MacSwiney of Mashanaglass

Sauce for the Mongoose
Bruce Kinloch

Argh: The Tale of a Tiger
M E Buckingham

Boxer and Beauty: A Tale of Two Cart-horses
Alfred Ollivant

A Chipmunk on my Shoulder
G J Helbemae

Grey Owl and the Beaver
Harper Cory